Ultimate car repair and maintenance

Your Complete Guide to Car Maintenance From Do-It-Yourself Skills to Become a Better Mechanic

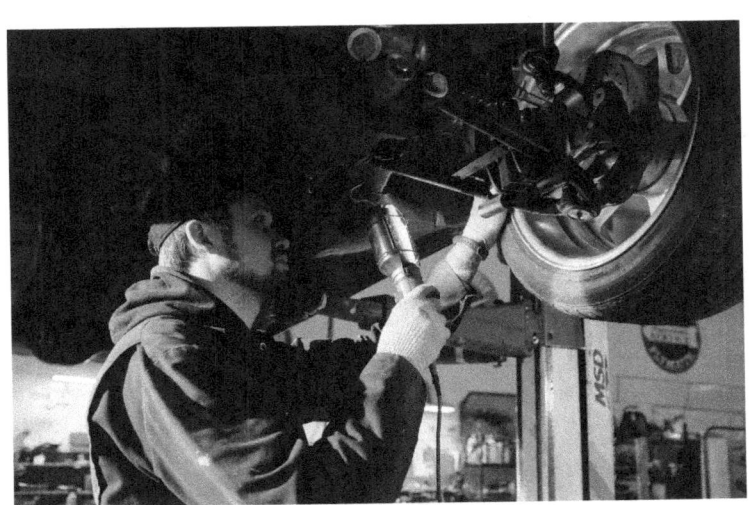

ROONIE D. KISSNER

Ultimate car repair and maintenance

ROONIE D. KISSNER

All rights reserved. No part of this publication may be reproduced, distributed, or transmitted in any form or by any means, including photocopying, recording, or other electronic or mechanical methods, without the prior written permission of the publisher, except in the case of brief quotations embodied in critical reviews and certain other noncommercial uses permitted by copyright law.

Copyright © Ronnie D. Kissner, 2024.

About The Author

Roonie D. Kissner is an African author and publisher who has made a name for himself in the nonfiction genre. Born and raised in Africa, Kissner developed a passion for reading and writing at an early age. He pursued his love for literature by applying for a book publishing training.

After completing this training, Kissner worked as a freelance writer and editor for various individuals. However, his desire to share knowledge and inspire others led him to start writing and publishing his own books.

Kissner's books cover a wide range of topics, from self-help and personal development to business and finance. He believes in providing practical advice and actionable steps that his readers can apply to their lives. His books are written in a clear and concise manner, making them accessible to readers of all backgrounds.

Kissner's works have received critical acclaim for their relevance and effectiveness. Many readers have reported that his books have helped them overcome challenges and achieve their goals. His dedication to empowering people through his writing has earned him a loyal following of readers who eagerly await his next release.

Today, Kissner continues to write and publish nonfiction books on Amazon KDP, sharing his knowledge and insights with readers around the world. He is a proud African author who is committed to making a positive impact on the lives of his readers through his writing..

Titles by Roonie D. Kissner

Daily Bible Devotional For Couples
The Ultimate Procrastination Cure
Designing Your Life
Be A Productive Small Business Owner
Car Repair And Maintenance.
Daily planner 2023
The Ultimate guide to Mindfulness

The Minimalist Cookbook.

Follow the author with this link https://www.amazon.com/author/roonied.kissner for more book updates.

Table of contents

	2
Table of contents	10
Introduction	13
Safety precautions and tools needed	17
Chapter 1: Understanding Your Vehicle	22
Vehicle anatomy and parts	22
How different systems work together	27
Chapter 2: Basic Maintenance	32
Oil changes and lubrication	32
Changing air filters	37
Tire care and maintenance	41
Chapter 3: Routine Inspections	47
Regular checks for fluid levels	47
Visual inspections for wear and tear	58
Chapter 4: DIY Repairs for Beginners	64
Changing spark plugs	64
Brake pad replacement	68
Headlight and taillight replacement	73
Chapter 5: Preventative Maintenance	78
Timing belt and water pump replacement	78
Transmission fluid change	83
Cooling system maintenance	87
Chapter 6: Troubleshooting Common Problems	92

Engine overheating	92
Strange noises and vibrations	98
Warning lights and what they mean	102
Chapter 7: Specialized Maintenance	**107**
Air conditioning maintenance and repair	107
Alignments and tire balancing	112
Handling bodywork and rust prevention	117
Changing a flat tire	123
Jump-starting a car	129
Chapter 9: Maintenance Schedules	**134**
Creating a personalized maintenance schedule	134
Tracking maintenance and repairs	139
Chapter 10: When to Call a Professional	**145**
Recognizing when a repair is beyond DIY	145
Finding a reputable mechanic	150

Introduction

Countless individuals rely on their automobiles on a daily basis to get them to their destinations. Our automobiles are reliable travel companions that we rely on for everything from the daily commute to work to long road trips across the nation or even just a quick trip to the grocery store. They provide us with independence, ease, and the excitement of traveling. But they need TLC to be healthy and happy, just like any pet.

In "Ultimate automobile Repair and Maintenance," you will find all the information you need to know about fixing and maintaining your automobile. Anyone from a complete DIY mechanic seeking to brush up on their skills to a complete newbie who has never turned a wrench is sure to find something useful in this comprehensive guide. We want to make it easy for you to know how to maintain your vehicle so that it runs safely, effectively, and without any problems.

Embark on a trip inside your vehicle's inner workings on the pages that follow. Let's begin at the beginning, with an explanation of your car's basic anatomy and how its many systems work together. Regular maintenance like oil changes and air filter replacements will become second nature to you.

As you continue, I'll dig into more complicated fixes and diagnostics, giving you the confidence to handle difficult problems and save on pricey mechanic fees. I'll describe your car's computer system and how to understand those weird warning lights on your dashboard.

Maintenance is not only about resolving issues; it's also about avoiding them. You'll explore the significance of preventive maintenance, including timing belt replacements and transmission fluid changes. I'll discuss typical automobile difficulties and show you how to diagnose and repair them.

No matter your degree of skill, this book has something to offer. I'll assist you through building a tailored maintenance program and monitoring your car's health over time. And for those occasions when an issue beyond your DIY talents, I'll advise you on hiring a reliable specialist.

In an era when environmental concerns are crucial, I'll also investigate sustainable and eco-friendly auto maintenance procedures to lessen your vehicle's ecological imprint.

As you read these pages, you'll start on a voyage of education, self-reliance, and care for your beloved four-wheeled buddy. Whether you're a weekend tinkerer, a responsible automobile owner, or just interested about the inner workings of these extraordinary machines, you'll discover significant insights inside this book.

Your automobile is more than simply a source of transportation; it's a representation of your independence and freedom. Join me on this adventure to become the ultimate caregiver of your beloved friend, and together, We'll guarantee that your vehicle delivers you securely and dependably wherever life's roads may lead.

Safety precautions and tools needed

Ensuring safety during automotive repair and maintenance is crucial. Alongside safety measures, having the correct equipment at your disposal is vital for a successful and secure maintenance routine. Let's go into the required safety measures and tools needed:

Safety Precautions:

1. Protective Gear: Always use proper protection gear, including safety glasses, gloves, and closed-toe shoes. If you're working below the vehicle, consider utilizing a creeper or a dependable jack stand for increased safety.

2. Ventilation: Work in a well-ventilated location, particularly if you're working with chemicals or exhaust fumes. Carbon monoxide is a quiet but deadly menace, therefore never operate the engine in an enclosed place.

3. Read the Manual: Before commencing any maintenance work, acquaint yourself with your car's handbook. It contains vital safety information and particular instructions pertaining to your vehicle.

4. Emergency Preparedness: Have a first aid kit readily accessible. Additionally, know the location of emergency exits and how to utilize fire extinguishers.

5. Secure Work Area: Ensure that your work place is clean, orderly, and well-lit. Remove any tripping risks and make sure your tools are in excellent condition.

6. Disconnect the Battery: When working on electrical components, disconnect the automobile battery to prevent the danger of electrical shock or damage to electronic components.

Tools Needed:

1. simple Hand Tools: A set of quality wrenches, screwdrivers, pliers, and a socket set are necessary for most simple maintenance chores.

2. Jack and Jack Stands: These are vital for safely lifting and securing your car while working beneath. Never depend entirely on a jack; always utilize jack stands for extra support.

3. Oil Filter Wrench and Oil Drain Pan: Essential for oil changes, these tools enable you remove and replace the oil filter and collect old oil.

4. Tire Pressure Gauge: Maintaining adequate tire pressure is critical for safety and fuel economy. A tire pressure gauge lets you monitor and adjust tire pressure.

5. Jumper Cables: Dead batteries happen. Jumper cables may be a lifeline and enable you to jump-start your automobile with the aid of another vehicle.

6. Flashlight: A decent quality flashlight is crucial, particularly if you're working in poorly lit places or at night.

7. Safety Glasses and Gloves: Protect your eyes and hands from any risks. Safety glasses screen your eyes from particles, while gloves offer a barrier against chemicals and sharp edges.

8. Torque Wrench: Precise torque is critical for many components. A torque wrench ensures that bolts are tightened to the manufacturer's

requirements, avoiding over-tightening or under-tightening.

Remember, safety should always be the primary concern. Taking the time to observe safety procedures and having the necessary equipment on hand will help to a smooth and secure auto repair experience.

Chapter 1: Understanding Your Vehicle

Vehicle anatomy and parts

Understanding the anatomy of your vehicle and its numerous components is important to being a professional car owner and DIY mechanic. Let's review the basic characteristics of vehicle anatomy and the functions of its primary components:

1. Chassis and Frame: - The chassis acts as the structural base of the vehicle, supporting the body and other components.
   - The frame offers stiffness and strength, vital for the overall stability of the vehicle.

2. Engine: - The engine is the powerhouse of the vehicle, transforming gasoline into mechanical

energy. - Components include the cylinder block, pistons, crankshaft, camshaft, and different belts and pulleys.

3. gearbox: - The gearbox, whether automatic or manual, distributes power from the engine to the wheels. - Components include the gearbox, clutch (in manuals), and torque converter (in automatics).

4. Exhaust System: - Responsible for releasing exhaust gasses from the engine. - Components include the exhaust manifold, catalytic converter, muffler, and tailpipe.

5. Suspension System: - Enhances ride comfort and vehicle handling. - Components include shocks, struts, springs, control arms, and sway bars.

6. Steering System: - Enables the driver to regulate the direction of the vehicle. - Components include

the steering wheel, steering column, power steering pump, and steering linkage.

7. Braking System: - Responsible for slowing down and halting the vehicle. - Components include brake discs, brake pads, calipers, and the brake master cylinder.

8. Electrical System: - Powers and controls electrical components. - Components include the battery, alternator, starting motor, and numerous sensors.

9. Fuel System: - Manages the storage and distribution of fuel to the engine. - Components include the gasoline tank, fuel pump, fuel injectors, and fuel filter.

10. Cooling System: - Regulates the engine temperature to avoid overheating. - Components include the radiator, water pump, thermostat, and cooling fan.

11. Ignition System: - Initiates combustion in the engine. - Components include the spark plugs, ignition coil, distributor (in earlier systems), and electronic control module.

12. Drivetrain: - Transfers power from the gearbox to the wheels. - Components include the driveshaft, differential, and axles.

13. Tires and Wheels: - The link between the vehicle and the road. - Components include tires, rims, and the tire pressure monitoring system (TPMS).

14. Body: - The outward and inner construction of the vehicle. - Components include doors, windows, roof, seats, and dashboard.

Understanding the roles and interaction of these components will equip you to do basic maintenance, detect difficulties, and communicate effectively with technicians. Whether you're a car aficionado or a casual driver, understanding your vehicle's anatomy increases your entire driving experience and adds to prudent auto ownership.

How different systems work together

A vehicle is a sophisticated machine including several components that work cooperatively to guarantee its flawless functioning. Understanding how these systems interact is vital for both automobile owners and aspiring DIY technicians. Let's study how various systems work inside a vehicle:

1. Powertrain Integration: - The powertrain, comprising of the engine and gearbox, is the heart of the vehicle.
    - The engine creates power, while the gearbox transmits it to the wheels, permitting movement.

2. Fuel and Ignition Systems: - The fuel system regulates the storage and distribution of fuel to the engine.

- The ignition system ignites combustion, producing the power required for the vehicle to move.

3. Cooling and Lubrication Systems: - The cooling system manages the engine temperature to avoid overheating.
   - The lubrication system ensures that moving elements inside the engine are well-lubricated, decreasing friction and wear.

4. Exhaust and Emission Systems: - The exhaust system expels combustion by-products from the engine. - The emission system, including the catalytic converter, helps minimize dangerous pollutants before releasing them into the environment.

5. Transmission and Drivetrain: - The transmission distributes power from the engine to the wheels.

- The drivetrain, comprising the driveshaft and axles, permits power transfer to guarantee smooth movement.

6. Suspension and Steering Systems:
   - The suspension system promotes ride comfort and stability.
   - The steering mechanism enables the driver to control the direction of the vehicle.

7. Braking System: - The braking system slows down and stops the vehicle.
   - Cooperative systems include the brake pedal, brake fluid, brake discs, and brake pads.

8. Electrical and Charging Systems: - The electrical system powers many components, including lighting, sensors, and entertainment systems.
   - The charging system, including the alternator, restores the battery's charge while the vehicle is operating.

9. Safety Systems: - Modern automobiles feature safety systems like ABS (Anti-lock Braking System), airbags, and traction control. - These technologies work together to increase safety when driving and in the case of a collision.

10. Computer and Sensors: - The vehicle's computer system (ECU) analyzes data from several sensors.

- Sensors monitor different characteristics, such as engine performance, emissions, and safety features, contributing to efficient vehicle operation.

11. Climate Control System: - The climate control system manages the interior temperature and airflow.

- It contains components like the air conditioning compressor, heating core, and ventilation system.

12. Tires and Wheels: - Tires and wheels are crucial to both the vehicle's mobility and its handling.

- The tire pressure monitoring system (TPMS) maintains optimal tire inflation for safety and efficiency.

Chapter 2: Basic Maintenance

Oil changes and lubrication

Performing frequent oil changes and lubrication is a crucial component of vehicle maintenance, adding to the lifetime and maximum performance of your automobile. Here's a step-by-step instruction on how to do these tasks:

Oil Change:

Materials Needed:

- New oil filter
- Engine oil of the specified viscosity and type
- Oil drain pan

- Wrenches (for removing the oil drain stopper and oil filter)
- Oil filter wrench - Funnel
- Rag or disposable gloves

Procedure:

1. Gather Materials: Park your car on a flat surface. Ensure you have all the essential supplies and tools.

2. Warm Up the Engine: Run the engine for a few minutes to warm the oil. Warm oil flows more readily and drags impurities out with it.

3. raise the Car: Use a car jack to raise the front of your vehicle. Make care to attach it with jack stands for safety.

4. find the Oil Drain Plug: Crawl under the car and find the oil drain plug on the oil pan.

5. Place the Oil Drain Pan: Position the oil drain pan behind the oil drain plug to capture the spent oil.

6. Remove the Drain Plug: Use a wrench to loosen and remove the oil drain plug. Allow the oil to drain entirely.

7. Remove and Replace the Oil Filter: Use an oil filter wrench to remove the old oil filter. Before installing the new oil filter, add a tiny quantity of oil to the rubber gasket on the top.

8. reinstall the Drain cap: Once the oil has drained entirely, reinstall the oil drain cap and fasten it firmly.

9. Add New Oil: Position a funnel over the oil filler cap and pour in the required quantity and kind of oil. Check your vehicle's handbook for specifics.

10. Check Oil Level: Start the engine and let it run for a minute. Turn it off and check the oil level using the dipstick. Add extra oil if required.

11. Dispose of spent Oil: Properly dispose of the spent oil at a recycling facility or an auto parts shop.

Lubrication:

<u>Materials Needed:</u>

- Grease gun - Lubricant suited for your vehicle's components
- Rag

<u>Procedure:</u>

1. Locate Fittings: Identify the grease fittings on your vehicle's chassis, suspension, and steering

components. Common fittings include those on ball joints, tie rod ends, and U-joints.

2. Clean Fittings: Use a towel to clean away dirt and debris from the grease fittings.

3. Load the Grease Gun: Load the grease gun with the proper lubricant.

4. Attach the Grease Gun: Attach the grease gun to the fitting and pump grease until you see new grease emerging. Repeat for all appropriate fittings.

5. Inspect Components: While greasing, examine components for any indications of damage or wear. Address any concerns swiftly.

6. Wipe Excess Grease: After lubrication, wipe away any excess grease to avoid gathering dirt.

7. Lower the Vehicle: If you elevated the automobile for greater access, gently lower it using the jack.

Performing oil changes and lubrication at regular intervals, as advised by your vehicle's handbook, is vital for preserving maximum performance and minimizing premature wear. Always stick to safety regulations, and if you're unclear or uncomfortable with these chores, seek a professional technician.

Changing air filters

Changing your vehicle's air filter is a simple but vital element of routine maintenance. The air filter serves a critical function in ensuring clean air enters the engine, which is necessary for combustion and overall engine efficiency. Here's a step-by-step instruction on how to replace your car's air filter:

Materials Needed:

- New air filter (compatible with your car)
- Screwdriver or socket set (for loosening fasteners)

Procedure:

1. Identify the Air Filter Location: - Refer to your vehicle's handbook to find the air filter housing. It's frequently within a plastic or metal housing toward the top of the engine.

2. Gather Necessary Tools: - Ensure you have the proper tools, often a screwdriver or a socket set.

3. Park Safely: - Park your car in a safe and well-lit location. Turn off the engine and let it cool down if it has been running.

4. Open the Hood: - Pull the hood release lever inside your car to open the hood. Secure the hood with the prop rod.

5. Locate the Air Filter Housing: - Find the air filter housing by referring to your vehicle's handbook or searching for a rectangular or round box near the engine.

6. Remove the bolts: - Use a screwdriver or socket set to remove the bolts fastening the air filter housing. These might be screws, clips, or nuts, depending on your car.

7. Open the Housing: - Once the fasteners are removed, open the air filter housing. Be mindful of any debris surrounding the housing to avoid it from entering the air intake.

8. Remove the Old Air Filter: - Take out the old air filter from the housing. Note the direction of the airflow marked on the side of the filter; this will help you install the new one appropriately.

9. Inspect the Filter: - Check the old air filter for dirt, debris, or damage. If it's very dusty or damaged, it's certainly time for a replacement.

10. Install the New Air Filter: - Place the new air filter into the housing, ensuring it fits tightly. Align it with the airflow direction specified on the side.

11. Close the Housing: - Close the air filter housing and secure it by tightening the bolts removed previously.

12. Close the Hood: - Gently lower the hood and secure it in place.

13. Start the Engine: - Start your car and let it run for a few minutes. This helps ensuring the new air filter is correctly seated.

Changing your air filter regularly—usually every 12,000 to 15,000 miles or as advised by your vehicle's manual—ensures your engine gets clean air, increasing fuel economy and performance. It's an easy activity that adds to the general health and durability of your car.

Tire care and maintenance

Spending only a few minutes each month on maintenance may make all the difference in your tires' safety, life and performance. You don't need to be an expert technician. Just stick to the four fundamentals.

1. TIRE PRESSURE

Proper inflation pressure—perhaps the most critical tire condition to monitor—gives tires the capacity to sustain the vehicle and you to regulate it for best performance. Bonus benefit: Maintaining optimum inflation pressure increases fuel efficiency, too.

Under inflated tires cause excessive heat build-up and tension, producing uneven wear and internal damage.
Over inflated tires are more likely to be cut, punctured or damaged when encountering an impediment, such as a pothole.

CHECKING YOUR PRESSURE

Use a tire gauge to check inflation pressure, measured in in PSI (pounds per square inch). You'll discover suggested pressure on a sticker on the driver's door or in your car owner's handbook.

And don't forget about your spare.

WHEN TO CHECK

Check your tires at least once a month, and check them while they're cold—meaning parked for at least three hours. Note that inflation pressure rises (in warm weather) or reduces (in cold weather) one to two pounds for every 10 degrees of temperature variation.

Also, check your pressure before you leave out on any lengthy travels, pull large items or tow a trailer.

TPMS

Tire Pressure Monitoring Systems (TPMS) detect loss of inflation pressure and notify drivers when tires are 25% under inflated. For many automobiles, this warning may be too late to avoid harm. TPMS are not a substitute for monthly tire pressure checks using a gauge.

2. TREAD DEPTH

Tread equals traction—giving your tires a hold on the road, particularly in inclement weather. Lose too much tread and you might lose control.

Once your tread wears down to 2/32nds of an inch, it's time to contact your local tire shop. Take a few minutes each month to physically examine your tires for uneven wear, high and low spots, abnormally smooth patches and other symptoms of degradation.

Tires contain built-in tread wear indicators, called wear bars. When the tops of these bars are flush with the tire's tread, the tire has to be changed.

Here's an easy and inexpensive tread test: Place a coin upside down into a tread groove. If part of Lincoln's head is covered by the tread, you're good to go. If you can see all of his head, it's new tire time.

3. TIRE ROTATION

Rotating your tires based on the guidelines in your vehicle's owner's handbook helps avoid uneven tire wear. If no rotation interval is given, USTMA advises every 5,000 to 8,000 miles.

Once you're done, ensure sure the inflation pressure is adjusted to the car manufacturer's standards.

4. TIRE ALIGNMENT

Striking a pothole or other road hazard might create alignment concerns. Misaligned wheels may lead to uneven, quick tread wear and should be repaired by a tire provider. Have your alignment examined at any hint of difficulty, such as "pulling," and frequently, along with your tire balance, as indicated by your vehicle's owner's handbook.

Chapter 3: Routine Inspections

Regular checks for fluid levels

Your automobile is a major investment, one that demands ongoing upkeep. Checking its different fluids periodically can assist to defend against failure, mechanical damage, and even avoidable accidents. Fortunately, learning to keep an eye on your vehicle's fluid levels is pretty straightforward, and doesn't take long if you know how to discover what you're looking for.

Part 1

1. Make it a point to check the fluids in your car generally every 4-6 months. Your owner's handbook will give you an idea of when you should have a glance at the fluid levels of each main component

beneath the hood. However, this duration is generally the minimal minimum frequency required to maintain your warranty in effect. A better rule of thumb is to check your fluids around twice a year, or every 5,000-10,000 miles (whichever comes first).

If you're the forgetful kind, it may be a good idea to mark your calendar or set a reminder on your gadget. Your car's fluids are its lifeline. Regular inspections will assist you ensure that it has all it needs to continue working cleanly and effectively.

2. Park your automobile on a smooth, level area and set the parking brake. Pull the brake handle up as far as it will go to make sure that the braking system is completely engaged. Setting the parking brake will keep your car from rolling or shifting suddenly while you're fooling about beneath the hood. If you have a button-style parking brake, just press it in all the way to activate it. The safest area to check your

car's fluids is within the garage or in a parking lot that isn't too crowded.

3. Pop your car's hood to access your fluid tanks. Look around the console area for a little hand lever that controls the locking mechanism for the hood. This lever is normally situated anywhere around the bottom half of the left hand side of the dashboard and labeled with the symbol of a vehicle with its hood up for easy identification.

When you discover it, pull it towards you. You'll hear an audible click when the hood lifts.
With some automobile types, it may be required to push a separate latch on the underside of the hood itself in order to open it up all the way.

Use the thin metal rod located to one side of the engine compartment to keep your hood propped up while you operate.

Part 2

Assessing Your Various Fluid Levels

The oil dipstick.

1. Start by checking your engine oil. Locate the yellow or white oil dipstick protruding from the top of the engine and hook your finger through the loop. Pull the dipstick all the way out, taking care to loosen any clips that could be keeping it in place. Use a paper towel or cloth to wipe the dipstick clean, then re-insert the dipstick into its opening and push it in as far as it will go. Pull the dipstick out again and examine the oil level. When you're done, secure the dipstick back in its opening.

Always check the oil after the automobile has had an hour or so to calm down. That manner, the oil in the return galleries, cylinder head valleys, and other components will have an opportunity to drain, avoiding misleading readings. The dipstick includes marks on it showing a range of permissible oil levels (typically notched, dimpled, or scribed). Double verify the marks you see to the diagrams in your owner's handbook. If the oil level is too low, you'll need to add a suitable motor oil straight soon.

Note the color of your oil, too. Clean engine oil has a transparent golden tint. Dirty engine oil will often be black or dark brown. If your oil seems unclean, verify your car's records to discover when the oil was last changed. A vehicle may operate on slightly darker oil just fine, so it's preferable to go by the timetable than the color alone.

Schedule your oil changes based on time rather than mileage alone. Even if you don't travel the prescribed amount of miles, it's a good idea to change your oil around once every six months, or more often if you do a lot of driving. It's possible for your car's oil to break down and become less efficient even if it's simply sitting in the garage.

Repeated, severe loss of engine oil might be a signal of a leak. Keep a careful check on the ground beneath the location where you regularly park to search for telltale oil streaks. If you notice any, get your car into a shop to have it looked at.
If your oil appears milky or foamy, it could be contaminated with coolant. This could point to a blown head gasket or another serious issue.

2. Take a look at your transmission fluid. For best results, do this while the engine is running and fully warmed up (either in neutral or park, depending on the make and model). It will be the second of two dipsticks on the engine, usually colored red. As you did with the oil dipstick, pull it out, wipe it off, push it all the way back in, then slide it out again and check the level. Once again, look for the fluid to

fall between the two notches, grooves, or marks on the dipstick.

Healthy transmission fluid will have a glossy reddish hue. If yours looks brown or black or has a distinct burnt odor, it's probably time to replace it.

Your transmission fluid doesn't need to be changed anywhere near as often as your engine oil. In newer cars, the recommended service interval may be as high as 100,000 miles (160,000 km). Consult your owner's manual to find more concrete guidelines for the model you drive.

This fluid is responsible for lubricating the transmission, or your car's gear system.

3. Inspect your brake fluid levels. Scan the engine compartment for a plastic reservoir labeled "brake fluid," or flip through your owner's manual to pinpoint its location. With most reservoirs, you'll be able to read the fluid level right through the plastic. Wipe any dirt, dust, or debris off the outside

of the tank, if necessary. If you still can't get a good look at the fluid, twist the cap off and peek inside.

If your brake fluid reservoir is especially hard to see through, it may help to jostle your vehicle gently on its suspension to get the fluid sloshing around and create visible movement.

Cars shouldn't consume brake fluid, no matter how old or hard-driven they are. If your brake fluid looks low, have your car checked to find out why. The culprit may be a leak in the brake line or worn brake surface, which in the worst case scenario could cause your vehicle to fail to stop.

4. Eyeball your power steering fluid. This will also ordinarily be contained in a plastic reservoir close to the belts on the passenger side of the vehicle. Read the level through the walls the same way you did the brake fluid. In some cases, there may be two pairs of lines: one for a hot engine and one for a cold engine. Check the one that's most appropriate for the state that your car is currently in.

If you need to add more power steering fluid, you can do so by screwing the lid off the reservoir and pouring in an appropriate product up to the indicated fill line.

These days, many cars are equipped with electronic power steering, which means they won't have a fluid reservoir.

5. Evaluate your coolant levels. The coolant is housed in a reservoir at the front of the engine compartment near the radiator. This is another one that you can read right through the plastic tank. Ideally, your coolant should be clear and its original color. If it's colorless, littered with small particles, or looks sludgy or gritty, it's likely contaminated, in which case you'll want to replace it ASAP.

Never check your coolant levels without first allowing your engine to cool down completely. Opening the reservoir while it's under pressure could cause scalding-hot water to come spraying out! Cars are designed to use antifreeze as coolant, not water. Antifreeze has a lower freezing point and a higher boiling point than water. If you need to replenish your coolant, make sure you're using the right stuff.

Be sure to read the label on the product you pick up. Some formulations can be added at full strength, while others must be mixed with an equal quantity of water. Sometimes there may be coolant in the reservoir but not in the radiator. If your reservoir is full but your car is running hot, remove the radiator hose to see if there's enough fluid in the radiator.

6. Top off your windshield wiper fluid if necessary. While low wiper fluid levels won't affect the performance of your car, they will affect your visibility, safety, and overall driving experience. To see how much wiper fluid you've got in reserve,

look for a brightly-colored container bearing the image of a windshield near the rear of the engine compartment. When you find it, lift the cap to check the contents. If needed, fill the container all the way to the top before pressing the cap back into place.

Specialty wiper fluids formulated to easily cut through bugs and other road grime tend to be inexpensive, so there's no excuse not to buy a quality product. However, adding a little water or window cleaner to the reservoir will also do the trick in a pinch. If you live somewhere with a cold climate, choose a type of fluid that won't freeze when temperatures drop. Wiper fluids with low freezing points will be plainly labeled as such.

Visual inspections for wear and tear

Regular visual inspections for wear and tear are a proactive approach to vehicle maintenance, enabling you to discover possible faults before they become severe problems. Here's a tip on how to visually check various components of your vehicle:

1. Tires:

Why It's Important:
Worn or damaged tires may impact safety and control.

Visual Inspection:
1. Check for uneven wear, bulges, or cuts on the tire sidewalls.
2. Measure tread depth; replace tires if the tread is below the acceptable limit.

3. Look for embedded items or symptoms of a puncture.

2. Brake Components:

Why It's Important: Brake difficulties might lead to safety risks.

Visual Inspection:
1. Inspect brake pads and discs via the wheel spokes.
2. Look for brake fluid leaks around the calipers and brake lines.
3. Check for uneven wear on brake pads.

3. Suspension System:

Why It's Important: A broken suspension impacts ride comfort and handling.

Visual Inspection:
1. Inspect shock absorbers and struts for leakage.

2. Check for apparent damage or corrosion on suspension components.

4. Exhaust System:

Why It's Important:
A faulty exhaust system may contribute to low fuel economy and pollutants.

Visual Inspection:
1. Look for corrosion or holes in the exhaust pipes and muffler.
2. Inspect the exhaust system for loose or damaged hangers.

5. Fluid Leaks:

Why It's Important:
Fluid leaks may signal concerns with numerous vehicle systems.

Visual Inspection:

1. Check beneath the car for oil, transmission fluid, coolant, or brake fluid leaks.
2. Inspect the engine bay for any obvious fluid leaks around hoses and connectors.

6. Belts and Chains:

Why It's Important:
Worn or loose belts may contribute to engine performance difficulties.

Visual Inspection:
1. Check the drive belts for cracks, fraying, or evidence of wear.
2. Inspect the timing belt or chain for any evident faults.

7. Lights and Indicators:

Why It's Important: Proper lighting is crucial for safety.

Visual Inspection:
1. Regularly check all external lights, including headlights, brake lights, turn signals, and license plate lights.
2. Inspect the condition of the light lenses for fractures or damage.

8. Battery:

Why It's Important:
A failed battery might lead to starting troubles.

Visual Inspection:
1. Check for obvious evidence of corrosion on the battery terminals.
2. Inspect the battery casing for any bulges or cracks.

9. Wipers and Washer Fluid:

Why It's Important: Clear sight is vital for safe driving.

Visual Inspection:
1. Inspect wiper blades for wear and tear.
2. Check the washer fluid reservoir and replenish if required.

10. Undercarriage:

Why It's Important:
The undercarriage might disclose concerns such as corrosion or damage.

Visual Inspection:
1. Inspect the undercarriage for rust, particularly in places prone to corrosion.
2. Look for any loose or dangling components.

Chapter 4: DIY Repairs for Beginners

Changing spark plugs

Changing spark plugs is an easy but crucial element of normal automobile maintenance. Spark plugs serve a critical function in the ignition system, producing the spark required to ignite the air-fuel combination in the engine cylinders. Here's a step-by-step instruction on how to replace spark plugs:

Materials Needed:

- New spark plugs (check to your car handbook for the required kind and gap)
- Spark plug socket and extension
- Ratchet - Spark plug gap tool

- Anti-seize lubricant (optional)
- Torque wrench (optional)

Procedure:

1. Gather Materials:
   - Park the car in a safe and well-lit place.
   - Ensure the engine is cold before starting.

2. Locate the Spark Plugs: - Refer to your vehicle's handbook to find the spark plugs. They are frequently attached to large cables or ignition coils.

3. Prepare the Area: - Clean the area surrounding the spark plugs to avoid dirt from going into the cylinders when you remove them.

4. Remove the Spark Plug Wires or Coils: - Carefully unplug the spark plug wires or coils from the spark plugs. If your car has separate coils, use a ratchet and socket to remove them.

5. Remove the Old Spark Plugs: - Use a spark plug socket and extension on your ratchet to loosen and remove the old spark plugs.

6. Inspect the Old Spark Plugs: - Examine the old spark plugs for indications of wear, such as fouling or deposits. This may give insights on the engine's condition.

7. examine the Spark Plug Gap: - Use a spark plug gap tool to examine the gap of the new spark plugs. Adjust the spacing as required, following your vehicle's requirements.

8. Apply Anti-Seize (Optional): - Applying a tiny quantity of anti-seize lubricant to the threads of the new spark plugs may make future removal simpler. Be careful not to misuse it.

9. Install the New Spark Plugs: - Hand-tighten the new spark plugs into the cylinder head. Once snug, use the spark plug socket and ratchet for the final tightening.

10. Reconnect Spark Plug Wires or Coils: - Reattach the spark plug wires or coils to the new spark plugs. Ensure a secure connection.

11. Repeat for All Cylinders: - Repeat the process for each cylinder, guaranteeing uniformity in the replacement method.

12. Torque (Optional): - If you have a torque wrench, you may use it to tighten the spark plugs to the appropriate torque level. Refer to your car handbook for this information.

13. Start the Engine: - Start the engine to check it operates smoothly. Listen for any strange noises or misfires.

Brake pad replacement

Replacing brake pads is an essential maintenance activity that protects the safety and maximum functioning of your vehicle's braking system. Here's a step-by-step instruction on how to change brake pads:

Materials Needed:

- New brake pads
- Brake fluid (if required)
- C-clamp or brake caliper tool
- Lug wrench or tire iron
- Jack and jack stands
- Wrench set - Socket set
- Screwdriver - Brake cleaner - Brake grease

Procedure:

1. Gather Materials:

- Park the car on a level surface and apply the parking brake.

- Ensure the engine is off and the car is in gear (or "Park" for automatic transmissions).

2. Loosen Lug Nuts: - Use a lug wrench or tire iron to gently loosen the lug nuts on the wheel matching to the brake pads you're replacing.

3. raise the car: - Use a jack to raise the car off the ground, then secure it using jack stands.

4. Remove the Wheel: - Completely remove the lug nuts and take off the wheel to access the braking components.

5. Locate the Brake Caliper: - Locate the brake caliper, which is normally secured in place by bolts.

6. Remove Caliper Bolts: - Use a wrench or socket set to remove the bolts connecting the caliper to the caliper bracket.

7. Retract the Caliper Piston: - Use a C-clamp or a brake caliper tool to retract the caliper piston. This creates room for the bigger, thicker brake pads.

8. Remove Old Brake Pads: - Slide out the old brake pads from the caliper bracket. Take note of how they are positioned for reference during reassembly.

9. Clean the Caliper Bracket: - Use a screwdriver or brake tool to clean the caliper bracket, removing any dirt or debris. Apply brake cleaner if required.

10. Install New Brake Pads: - Slide the new brake pads into the caliper bracket, ensuring they are positioned the same manner as the previous ones.

11. Reattach the Caliper: - Slide the caliper back over the brake pads and secure it in place by tightening the caliper nuts.

12. Repeat for Other Brake Pads: - Repeat the method for any more brake pads you're replacing.

13. Replace the Wheel: - Put the wheel back on and hand-tighten the lug nuts.

14. Lower the car: - Use the jack to lower the car to the ground.

15. Tighten Lug Nuts: - Use a lug wrench or tire iron to firmly tighten the lug nuts in a crisscross pattern.

16. Brake Bedding (Optional): - After changing brake pads, it's advised to undertake a brake bedding operation to guarantee appropriate contact and performance.

17. Check Brake Fluid Level: - If required, check the brake fluid level in the master cylinder and fill it up with the appropriate brake fluid.

<u>Note:</u> Consult your vehicle handbook for detailed directions and torque requirements.
- Replace both sets of brake pads on an axle even if only one seems worn.
- Be careful not to over-tighten lug nuts, and follow the prescribed torque parameters.

If you're hesitant or uncomfortable with brake pad replacement, seek the aid of a professional technician. Properly working brake pads are crucial for safe driving, and frequent inspections and timely replacements add to the general well-being of your car

Headlight and taillight replacement

Replacing headlights and taillights is an easy but vital component of car maintenance, maintaining visibility and safety on the road. Here's a guide on how to change headlights and taillights:

Materials Needed:
- Replacement headlights or taillights (refer to your car handbook for the relevant bulb types)
- Screwdriver or socket set (for certain cars)
- Latex gloves or a clean towel (to prevent touching the bulb)
- Optional: bulb grease or dielectric grease

Procedure:

1. Gather Materials:
   - Park the car on a level surface, switch off the engine, and activate the parking brake.

- Ensure the headlights or taillights are cool before starting.

2. Locate the Bulb Holder: - For headlights, reach the bulb holder via the engine compartment.

 - For taillights, reach the bulb holder via the trunk or rear panel, or occasionally from the outside.

3. Remove the Old Bulb: - Depending on your car, you may need to remove a cover or use a screwdriver to reach the bulb holder.

 - For most bulbs, turn the socket counterclockwise to loosen it, and then gently remove the old bulb.

4. Install the New Bulb: - If your new bulb has a protective cover, remove it before installation.

 - Carefully place the new bulb into the socket. Avoid touching the glass component of the bulb with your bare hands, since oils from your skin might harm it.

5. put the Bulb: - Turn the socket clockwise to put the new bulb in place.

6. Optional: Apply Bulb Grease: - Applying a small quantity of bulb grease or dielectric grease to the base of the bulb may assist prevent corrosion and guarantee a strong electrical connection.

7. Test the Lights: - Turn on the headlights or taillights to confirm the new bulbs are operating appropriately.

8. Repeat for Other Bulbs: - If you're changing many bulbs, repeat the method for each one.

9. Secure the Cover (if applicable): - If you have to remove a cover or panel, secure it back in place.

Check your Vehicle Manual: Different cars may have distinct techniques for accessing and changing bulbs, so check your vehicle handbook for detailed instructions.

- Wear Gloves: If you're handling the bulb directly, use latex gloves or use a clean towel to prevent contacting the glass.

For Halogen Bulbs (sealed beam):

1. Locate the Bulb Holder: - For sealed beam headlights, the complete headlight component may need to be changed.
   - Access the assembly via the engine compartment.

2. Remove the Old Assembly: - Unscrew the retaining screws or nuts holding the headlight assembly.
   - Carefully remove the old assembly.

3. Install the New Assembly: - Align the new headlight assembly with the mounting points.

- Secure it in place by tightening the retaining screws or bolts.

4. Test the Lights: - Turn on the headlights to confirm the new assembly is operating appropriately.

5. Repeat for opposite Headlight: - If you're changing both headlights, repeat the method for the opposite side.

Replacing headlights and taillights is a chore most automobile owners can do. However, if you're uncertain or uneasy, it's always best to visit a skilled technician. Regularly testing and changing bulbs add to the safety and visibility of your car on the road.

Chapter 5: Preventative Maintenance

Timing belt and water pump replacement

Replacing the timing belt and water pump is a critical maintenance procedure, preventing potential engine damage and ensuring the longevity of your vehicle. Here's a step-by-step instruction on how to repair the timing belt and water pump:

Materials Needed:

- Replacement timing belt
- Replacement water pump
- Tensioner and idler pulleys (if not sold with the timing belt kit)
- Coolant

- Serpentine belt (if necessary)
- Gaskets or seals (if not included with the water pump)
- Socket set and wrenches
- Screwdrivers
- Harmonic balancer puller (if needed)
- Torque wrench
- Floor jack and jack stands

Procedure:

1. Gather Materials:
- Park the car in a well-ventilated and safe location.
- Disconnect the battery.

2. Locate the Timing Belt Components:
- Refer to your vehicle handbook to find the timing belt cover, tensioner, idler pulleys, and the water pump.

3. Loosen the Drive Belts (if necessary):

- If the timing belt is powered by the serpentine belt, loosen the serpentine belt to access the timing belt components.

4. Remove Timing Belt Cover:
- Remove the bolts securing the timing belt cover and take it off.

5. Align Timing Marks:
- Rotate the engine to align the timing markings on the crankshaft and camshaft(s) as recommended in your vehicle handbook.

6. Release Tension on Timing Belt:
- Loosen the tensioner to release tension on the timing belt.

7. Remove Timing Belt:
- Carefully slide off the timing belt from the sprockets.

8. Inspect Tensioner and Idler Pulleys:
- Check the condition of the tensioner and idler pulleys. Replace them if there are signs of wear or damage.

9. Remove Water Pump:
- If the water pump is driven by the timing belt, remove the bolts securing it and take out the water pump.

10. Clean the Surfaces:
- Clean the surfaces where the new water pump and timing belt will be installed.

11. Install New Water Pump:
- Install the new water pump, ensuring that gaskets or seals are in place.

12. Install New Timing Belt:

- Carefully thread the new timing belt over the sprockets, following the appropriate route as per the timing markings.

13. Tension the Timing Belt:
- Apply tension to the timing belt by adjusting the tensioner as specified in your vehicle manual.

14. Rotate Engine:
- Rotate the engine by hand to confirm that the timing belt is correctly aligned.

15. Torque Bolts to Specifications:
- Torque the bolts for the water pump, tensioner, and idler pulleys to the manufacturer's recommendations.

16. Reinstall Timing Belt Cover:
- Place the timing belt cover back in place and fasten it with the bolts.

17. Reinstall Drive Belts (if loosened):
- If you loosened the serpentine belt, reinstall and tension it according to specifications.

18. Fill with Coolant:
- Refill the cooling system with the appropriate coolant.

19. Reconnect Battery:
- Reconnect the battery.

20. Start Engine and Check for Leaks:
- Start the engine and check for any coolant or oil leaks. Monitor the temperature gauge.

Transmission fluid change

To keep your vehicle's gearbox running smoothly and for as long as possible, changing the fluid is a necessary maintenance item. The procedure for changing the transmission fluid is as follows:

Materials Required:

- New transmission fluid (to find out the kind to use, see your handbook)
- A dish or pan to catch oil or other fluids
wrenches and a set of sockets
- A channel
- Goggles and a pair of rubber gloves
As a pair, Jack and Jack are
- Screwdriver or ratchet for removing the drain plug
- New transmission fluid filter (if applicable)

Procedure:

1. Gather Materials:
   - Park the car on a level surface and apply the parking brake.
   - Ensure the engine is off.

2. find the Transmission Drain Plug: - Refer to your car handbook to find the transmission drain plug. It's commonly located on the transmission pan near the bottom of the transmission.

3. raise the car: - Use a jack to raise the front of the car, then secure it with jack stands.

4. Place Oil Pan: - Position an oil pan or container underneath the transmission drain plug to capture the fluid.

5. Remove the Drain Plug: - Use a socket or wrench to loosen and remove the transmission drain plug. Allow the liquids to drain entirely.

6. Replace the Drain Plug: - Once the fluid has emptied, replace and tighten the drain plug firmly.

7. Locate the Transmission Fluid Fill Cap: - Locate the transmission fluid fill cap, normally situated near the engine. It could have a dipstick connected.

8. Remove the Fill Cap: - Remove the transmission fluid fill cap to enable air to enter while you drain the fluid.

9. Use a Funnel to Add fresh Fluid: - Insert a funnel into the transmission fill hole and begin adding the fresh transmission fluid. Refer to your vehicle handbook for the appropriate fluid volume.

10. Check Fluid Level: - With the engine running and the vehicle in "Park" or "Neutral," check the transmission fluid level using the dipstick. Add extra fluid if required.

11. Optional: Change Transmission Filter (if applicable): - Some cars have an internal transmission filter that may require replacement. If

your car has one, follow the manufacturer's suggestions.

12. Lower the car: - Carefully lower the car using the jack.

13. Check for Leaks: - Start the engine and let it run for a few minutes. Check for any evidence of leakage around the drain plug and transmission pan.

Cooling system maintenance

Maintaining your vehicle's cooling system is vital to avoid overheating, engine damage, and other associated concerns. Here's a guide on cooling system maintenance:

Materials Needed:

- Coolant (antifreeze)
- Distilled water
- Coolant tester or refractometer
- Funnel
- Hose and clamp inspection
- Radiator cover (if required)
- Coolant system cleanser (optional)
- Water pump inspection tools (if applicable)

Procedure:

1. Gather Materials:
    - Ensure the car is parked on a flat area and the engine is cool.
    - Engage the parking brake.

2. Check Coolant Level: - Open the hood and find the coolant reservoir. Check the coolant level; it should be between the "Min" and "Max" markers.

3. Inspect Hoses and Clamps:

   - Visually examine any coolant hoses for evidence of wear, leaks, or damage.

   - Check hose clamps for tightness and corrosion.

4. Pressure Test the Cooling System: - Perform a pressure test to check for leaks. Use a cooling system pressure tester and follow the manufacturer's instructions.

5. Check the Radiator Cap: - Inspect the radiator cap for any indications of wear, corrosion, or deterioration. Replace it if required.

6. Coolant Flush (if needed): - Refer to your vehicle handbook for suggested coolant changing intervals.

   - If the coolant is due for replacement, try doing a coolant cleanse. Follow the manufacturer's directions or use a coolant system cleaning.

7. Mix Coolant and Water: - If you're adding coolant, mix it with distilled water according to the manufacturer's guidelines. Use a 50/50 ratio for most climates.

8. Top Up the Coolant: - Add coolant to the coolant reservoir. Use a funnel to prevent spilling.

9. Bleed the Air from the System (if applicable): - Some cars may have air pockets following a coolant replacement. Refer to your car handbook for details on how to bleed the air from the cooling system.

10. Check for Leaks: - Start the engine and observe for any symptoms of coolant leakage. Address any leaks swiftly.

11. Check the Thermostat: - If your car is ready for a coolant change, it's a good opportunity to examine or replace the thermostat. Follow your car handbook for instructions.

12. examine the Water Pump: - If your vehicle has a visible water pump, examine it for leaks or strange sounds. Replace the water pump if required.

13. Test Coolant Mixture: - Use a coolant tester or refractometer to assess the concentration of coolant in the system. Ensure it offers suitable freeze and boil protection.

Chapter 6: Troubleshooting Common Problems

Engine overheating

Causes of Engine Overheating:

1. Low Coolant Level: - Insufficient coolant might lead to poor heat dissipation.

2. Coolant Leaks: - Leaks in the radiator, hoses, water pump, or other components might result in a loss of coolant.

3. Faulty Thermostat: - A faulty thermostat may cause the engine to run excessively hot or too cold.

4. Broken Water Pump: - A faulty water pump might hinder the passage of coolant.

5. Radiator Issues: - Clogged or damaged radiators inhibit adequate heat exchange.

6. Cooling Fan Problems: - Malfunctioning electric fans or a defective fan clutch might interrupt airflow through the radiator.

7. Blocked Coolant passageways: - Deposits or debris clogging coolant passageways might hinder flow.

8. Faulty Radiator Cap: - A broken or incorrectly functioning radiator cap may not retain sufficient pressure.

9. Excessive Engine Load: - Overloading the engine, such as pulling large weights, might cause overheating.

10. Faulty Head Gasket: - A burst head gasket might lead to coolant entering the combustion chamber.

Symptoms of Engine Overheating:

1. Temperature Gauge Warning: - The temperature gauge on the dashboard indicates higher-than-normal values.

2. Steam or Smoke: - Visible steam or smoke coming from the engine compartment.

3. Coolant Leak Puddles: - Puddles of coolant beneath the car.

4. Burning Smell: - A significant burning smell, suggesting overheated components.

5. Engine Misfire: - Overheating might cause the engine to misfire or operate roughly.

6. Reduced Power: - Engine performance may reduce as a preventative measure.

Possible Solutions:

1. Stop Driving Immediately: - If you observe indications of overheating, pull over, switch off the engine, and let it cool down.

2. Check Coolant Level: - Inspect the coolant reservoir and replenish up if required. Look for leaks.

3. Inspect Hoses and Radiator: - Check for apparent leaks or damage in hoses and the radiator. Replace as required.

4. Check the Thermostat: - Verify the thermostat is operating appropriately. Replace if required.

5. Check the Water Pump: - Inspect the water pump for leaks or strange sounds. Replace if required.

6. Check the Radiator Cap: - Ensure the radiator cap maintains correct pressure. Replace if damaged.

7. Flush the Cooling System: - Perform a coolant flush if the coolant is old or polluted.

8. Check the Electric Fans: - Inspect electric fans for appropriate functioning. Replace defective fans or the fan clutch.

9. Check the Radiator for Clogs: - If the radiator is blocked, consider flushing or replacing it.

10. Check for Head Gasket Issues: - Look for indicators of a blown head gasket, such as coolant in the oil or white smoke from the exhaust. Consult a professional if suspected.

11. Reduce Engine Load: - Avoid overloading the engine, particularly under harsh situations.

12. Consult a Professional: - If you're unable to identify or address the problem, seek the aid of a professional mechanic.

Preventive Measures:

- Regular Maintenance: - Follow the specified maintenance plan, including coolant changes.

- Inspect Hoses and Belts: - Regularly examine for wear or damage and replace as required.

- Keep Radiator Clean: - Remove debris from the radiator to promote effective heat exchange.

- Avoid Overloading: - Avoid exceeding your vehicle's towing or load capability.

- Address faults swiftly: - Address any cooling system faults swiftly to avoid additional harm.

Strange noises and vibrations

Strange Noises and Vibrations in a Vehicle: Causes and Solutions

Unusual sounds and vibrations in a car might signal underlying problems that need addressing. Here's a guide to assist discover probable reasons and solutions:

Common Causes of Strange Noises:

1. Squealing or Screeching: - Causes: Worn-out or loose serpentine belts, broken pulleys, or a faulty belt tensioner.

- Solution: Inspect and replace worn-out belts, tighten loose belts, and repair broken pulleys or the tensioner.

2. Grinding or Growling: - Causes: Worn brake pads, broken wheel bearings, or troubles with the gearbox.

  - Solution: Replace worn brake pads, examine and replace broken wheel bearings, and see a professional for gearbox difficulties.

3. Clicking or Ticking: - Causes: Faulty CV joints, worn-out lifters, or difficulties with the valve train.

  - Solution: Replace CV joints if broken, adjust or replace lifters, and handle valve train concerns as appropriate.

4. Knocking or Pinging: - Causes: Detonation in the engine, worn-out bearings, or troubles with the fuel system.

- Solution: Use the proper fuel grade, examine and replace worn bearings, and handle fuel system concerns.

5. Rattling or Clunking: - Causes: Loose or broken exhaust components, worn suspension parts, or faults with the steering system.

- Solution: Tighten or replace loose exhaust components, examine and replace worn suspension elements, and solve steering system concerns.

6. Hissing or Sizzling: - Causes: Coolant or oil leaks, difficulties with the exhaust system, or problems with the cooling system.

- Solution: Identify and rectify coolant or oil leaks, examine and repair the exhaust system, and handle cooling system faults.

Common Causes of Vibrations:

1. Unbalanced Tires: - Causes: Uneven tire wear, uneven tire pressure, or broken tires.
   - Solution: Rotate and balance tires, maintain adequate tire pressure, and replace damaged tires.

2. Wheel Alignment Issues: - Causes: Misaligned wheels due to striking potholes or curbs.
   - Solution: Get a professional wheel alignment to solve alignment difficulties.

3. Worn-out Brake Rotors: - Causes: Unevenly worn brake rotors.
   - Solution: Replace worn brake rotors and ensure correct brake pad installation.

4. Faulty Suspension Components: - Causes: Worn-out shocks, struts, or other suspension elements.

- Solution: Inspect and replace worn suspension components.

5. Issues with the Drivetrain: - Causes: Problems with the transmission, driveshaft, or differential.
 - Solution: Consult a technician to diagnose and treat drivetrain faults.

6. Engine Misfire: - Causes: Faulty spark plugs, ignition coils, or fuel injectors.
 - Solution: Replace spark plugs, ignition coils, or fuel injectors as required.

Warning lights and what they mean

Modern automobiles are equipped with numerous warning lights on the dashboard to notify drivers about possible concerns. Understanding these lights may help you solve issues swiftly. Here's a guide to common warning lights and what they mean:

1. Check Engine Light: - Meaning: Indicates a problem with the engine or emissions system.
 - Action: Schedule a diagnostic check to detect and resolve the problem.

2. Battery Warning Light: - Meaning: Indicates a problem with the charging system.
 - Action: Check the alternator belt and battery terminals. If troubles continue, visit a technician.

3. Oil Pressure Warning Light: - Meaning: Indicates low oil pressure.
 - Action: Stop the car immediately and check oil levels. If oil levels are adequate, seek expert treatment.

4. braking System Warning Light: - Meaning: Indicates faults with the braking system, such as low brake fluid or a brake system failure.

- Action: Check brake fluid levels. If low, top up. If the light remains, visit a technician.

5. ABS Warning Light: - Meaning: Indicates a fault with the Anti-Lock Braking System (ABS).

- Action: Continue driving but seek expert help to identify and remedy the problem.

6. Traction Control System (TCS) Light: - Meaning: Alerts to difficulties with the Traction Control System.

- Action: Seek expert examination and repair.

7. Airbag Warning Light: - Meaning: Indicates a problem with the airbag system.

- Action: Consult a mechanic to identify and solve the problem.

8. TPMS (Tire Pressure Monitoring System) Light: - Meaning: Indicates low tire pressure in one or more tires.

- Action: Check and adjust tire pressure. If the light remains, look for punctures or leaks.

9. Coolant Temperature Warning Light: - Meaning: Indicates the engine is overheating.

- Action: Stop the car, let the engine cool, and check coolant levels. If the issue continues, contact a technician.

10. Transmission Temperature Warning Light: - Meaning: Indicates excessive transmission fluid temperature.

- Action: Stop the car, let it cool, and check transmission fluid levels. Seek expert treatment if the situation continues.

11. ESP/BAS Warning Light: - Meaning: Indicates difficulties with the Electronic Stability Program (ESP) or Brake Assist System (BAS).

- Action: Seek expert examination and repair.

12. Glow Plug Indicator (Diesel Vehicles): - Meaning: Indicates the need to wait before starting the engine in a diesel vehicle.

- Action: Wait for the glow plugs to warm up before starting the engine.

Chapter 7: Specialized Maintenance

Air conditioning maintenance and repair

Maintaining and fixing your vehicle's air conditioning (AC) system is vital for comfort during hot weather. Here's a guide on AC maintenance and frequent repair tasks:

AC Maintenance Tips:

1. Regular Inspection: - Action: Visually examine the AC components for leaks, damage, or worn-out parts.

2. Check Refrigerant Levels: - Action: Ensure refrigerant levels are adequate. If low, it may suggest a leak.

3. Clean or Replace Air Filters: - Action: Regularly clean or replace the cabin air filter to guarantee appropriate airflow.

4. Inspect Belts and Hoses: - Action: Check for cracks or evidence of wear on belts and hoses. Replace if required.

5. Clean Condenser Coils: - Action: Remove junk from the condenser coils using a soft brush or compressed air.

6. Check Thermostat Calibration: - Action: Verify that the thermostat is appropriately calibrated for temperature control.

7. Run the AC System Regularly: - Action: Even during the winter, run the AC system for a few minutes to keep it lubricated and prevent seals from drying out.

Common AC Repair Tasks:

1. Refrigerant Leak Repair: - Symptoms: Reduced cooling efficiency, hissing noises.
 - Action: Identify and fix leaks, then recharge the system with the necessary refrigerant.

2. Compressor Replacement: - Symptoms: Lack of cold air despite correct refrigerant levels.
 - Action: Replace a defective compressor to restore cooling efficiency.

3. Condenser or Evaporator Coil Replacement: - Symptoms: Inadequate cooling.
 - Action: Replace broken or blocked coils to optimize AC performance.

4. Thermostat Replacement: - Symptoms: Inconsistent or imprecise temperature control.

- Action: Replace a defective thermostat to guarantee exact temperature management.

5. Blower Motor Repair or Replacement: - Symptoms: Weak airflow or no air.

- Action: Repair or replace the blower motor to restore appropriate airflow.

6. Expansion Valve Replacement: - Symptoms: Insufficient cooling, unstable temperature regulation.

- Action: Replace a malfunctioning expansion valve for enhanced AC performance.

7. Electrical System Inspection: - Symptoms: AC not turning on.

- Action: Check fuses, relays, and electrical connections. Replace or repair as required.

8. Condenser Fan Repair or Replacement: - Symptoms: Overheating, insufficient cooling.

- Action: Repair or replace a defective condenser fan.

Preventive Measures:

- Use the AC System Regularly: - Running the AC system frequently helps preserve its functioning.

- Prompt Attention to faults: - Address any AC faults swiftly to avoid additional harm.

- Professional Inspection: - Schedule frequent professional inspections to discover any concerns early.

- Use the Correct Refrigerant: - Ensure the use of the manufacturer-recommended refrigerant during recharges.

Note:

While certain AC maintenance activities may be done by car owners, many repairs, particularly those requiring refrigerant management, may need professional knowledge. If you're confused about a repair work or face difficult complications, seek the aid of a competent automotive expert. Regular maintenance and timely repairs contribute to the best operation and durability of your vehicle's air conditioning system.

Alignments and tire balancing

Wheel alignments and tire balancing are vital maintenance procedures that contribute to a smooth ride, even tire wear, and optimum vehicle handling. Here's a guide on completing wheel alignments and tire balancing:

1. Wheel Alignment:

Wheel alignment entails altering the angles of the wheels to ensure they are parallel to one other and perpendicular to the ground. Proper alignment reduces uneven tire wear and helps maintain straight driving.

Signs Your Vehicle Needs an Alignment: - Uneven tire wear.
- Vehicle pulling to one side.
- Steering wheel off-center.

Performing a DIY Inspection:

1. Tire Inspection: - Check tires for uneven wear, which might indicate alignment difficulties.

2. Steering Wheel Inspection: - While driving on a straight, level route, examine whether the steering wheel is centered.

DIY Alignment Check: Toe-In/Toe-Out: - Toe-In: Front of the tires closer together than the rear.
- Toe-Out: Front of the tires further apart than the rear.

Professional Alignment: - Toe, Camber, and Caster Adjustments: - These are the three major alignment angles modified by specialists.

How Often to Perform Wheel Alignment: - Recommendation: Check alignment yearly or if you observe indicators of misalignment.

2. Tire Balancing:

Meaning:
Tire balancing includes transferring weight throughout a tire and wheel assembly to provide uniform tire wear and a smoother ride. It solves

unequal weight distribution that may produce vibrations.

Signs Your Tires Need Balancing: - Vibrations in the steering wheel or vehicle.
- Uneven tire wear.

Performing a DIY Inspection: - Visual Inspection: - Check for evidence of uneven tire wear.

DIY Balancing Check: Static and Dynamic Balancing: - Static Balancing: Involves assessing the balance of the tire and wheel assembly when stationary.
- Dynamic Balancing: Involves testing the balance as the wheel turns.

Professional Tire Balancing: - utilize of a Balancing Machine: - Professionals utilize specialized equipment to diagnose and fix imbalances.

How Often to Perform Tire Balancing: - Recommendation: Balancing is normally advised every 5,000 to 6,000 miles or when you observe indicators of imbalance.

Performing DIY Wheel Alignment:

1. Use a String Method: - Secure a string around the sidewalls of the vehicle at the same height on both sides. Compare the distance between the string and the tires to check for alignment.

2. DIY Toe-In/Toe-Out Check: - Use a measuring tape to measure the distance between the front and rear of the tires on each side.

Performing DIY Tire Balancing:

1. Use a Bubble Balancer: - Place the wheel on a bubble balancer to determine heavy regions.

2. Dynamic Balancing using Weights: - Attach weights to the wheel rim at certain positions to establish dynamic balance.

Handling bodywork and rust prevention

Maintaining your vehicle's bodywork and avoiding rust are vital for sustaining its beauty and structural integrity. Here's a complete guide on managing bodywork and rust prevention:

1. Handling Bodywork:

1.1. Surface Scratches and Minor Damage:

- Light Scratches: - Action: Use a scratch removal kit to polish and eliminate the appearance of light scratches.

- Touch-Up Paint: - Action: For tiny chips or scratches, use a touch-up paint kit matching your vehicle's color.

1.2. Dents and Dings:

- DIY Dent Removal Kits: - Action: Use suction or leverage-based dent removal kits for tiny, accessible dents.

- Professional Repair: - Action: For bigger or intricate dents, call a professional vehicle body shop for repair.

1.3. Rust Spots:

- Remove Surface Rust: - Action: Sand the afflicted area, use a rust converter, and repaint.

- Cut Out and Repair: - Action: For deeper rust, cut off the afflicted area, treat with a rust converter, then weld in a new piece.

1.4. Paint Protection:

- Waxing and Sealants: - Action: Regularly wax your car to preserve the paint from environmental influences.

- Clear Coat Sealants: - Action: Apply clear coat sealants for an extra layer of protection.

2. Rust Prevention:

2.1. Regular Cleaning:

- Action: Clean your car periodically, giving attention to areas prone to dirt and debris collection.

2.2. Undercoating:

- Action: Apply an undercoating to protect the undercarriage from salt, moisture, and road debris.

2.3. Rust Inhibitors:

- Action: Use rust inhibitors, which come in spray or gel form, on sensitive locations such as wheel wells and joints.

2.4. Rust-Resistant Materials:

- Action: When acquiring a new car, choose models with rust-resistant materials in essential sections.

2.5. Keep Drains Clear:

- Action: Ensure that drainage channels, particularly around doors and windows, are clear to avoid water collection.

2.6. Rustproofing Treatments:

- Action: Invest in expert rustproofing treatments for total protection.

2.7. Garage Parking:

- Action: Park your car in a garage or covered location to shelter it from the weather.

3. General Tips:

- Regular Inspections: - Action: Periodically examine your car for indications of corrosion, dents, or scratches.

- Prompt Repairs: - Action: Address any damage or rust concerns swiftly to prevent them from spreading.

- Professional Assistance: - Action: For severe bodywork or intricate rust concerns, seek the skills of a professional car body shop.

Chapter 8: Emergency Repairs and Roadside Assistance

Changing a flat tire

Changing a flat tire is a fundamental skill any motorist should know. Here's a step-by-step instruction to assist you safely and effectively repair a flat tire:

1. Find a Safe Location:

- Action: Move your car to a secure and level position away from traffic. Turn on warning lights and apply the parking brake.

2. Gather Necessary Tools:

- Required Tools:

   - Spare tire - Jack - Lug wrench - Vehicle owner's manual

   3. Locate Spare Tire and Tools:

- Action: Identify the spare tire, jack, and lug wrench. These are normally positioned in the trunk or behind the floor mat in the cargo area.

   4. Loosen Lug Nuts:

- Action: Before moving the car, use the lug wrench to release the lug nuts on the flat tire. Do this while the car is still on the ground to prevent the wheel from spinning.

   5. Position the Jack:

- Action: Consult your car owner's handbook to identify the necessary jacking locations. Place the jack in the right place.

6. Lift the Vehicle:

- Action: Use the jack to elevate the car until the flat tire is roughly six inches off the ground. Avoid working beneath the vehicle.

7. Remove Lug Nuts and Flat Tire:

- Action: Fully remove the loosened lug nuts and take off the flat tire. Place lug nuts in a secure position.

8. Install Spare Tire:

- Action: Carefully insert the spare tire onto the wheel studs. Hand-tighten the lug nuts.

9. Lower the Vehicle:

- Action: Use the jack to lower the car until the spare tire makes contact with the ground.

10. Tighten Lug Nuts:

- Action: Use the lug wrench to tighten the lug nuts as much as possible in a crisscross pattern.

11. Lower Vehicle Completely:

- Action: Lower the car to the ground using the jack.

12. Final Lug Nut Tightening:

- Action: Once the car is on the ground, use the lug wrench to further tighten the lug nuts in a crisscross pattern.

13. Secure Tools and Flat Tire:

- Action: Place the flat tire, jack, and lug wrench back in the car. Ensure all tools are secure.

14. Check Spare Tire Pressure:

- Action: Before driving, check the spare tire's pressure with a tire pressure gauge. Inflate as required.

15. Visit a Tire Professional:

- Action: Replace the spare tire with a permanent solution as soon as feasible. Spare tires are meant for temporary usage.

Tips:

- Practice Changing a Tire: - Familiarize yourself with the tire-changing procedure in a controlled setting before an emergency.

- Regularly Check Spare Tire: - Ensure your spare tire is in excellent shape and properly inflated.

- get Professional Assistance if Unsure: - If you face problems or if situations are dangerous, get professional aid.

By following these instructions and practicing changing a tire, you'll be more prepared to manage this typical roadside issue. Remember that safety is the first concern, so always pick a safe site and practice care during the procedure.

Jump-starting a car

Here's a step-by-step tutorial to safely jump-start a car:

1. Ensure Safety:

- Action: Park the operating car near enough to the dead one so that the jumper wires may reach. Ensure both cars are in Park (for automatic) or Neutral (for manual), and switch off the ignition.

2. Inspect the Batteries:

- Action: Check both batteries for evidence of damage or corrosion. If the batteries are leaking or broken, do not continue with jump-starting.

3. Get Jumper Cables:

- Required Tools: - Jumper cables

4. Connect the Jumper Cables:

- Step 1: Connect Positive (Red) Cable: - Action: Connect one end of the red jumper wire to the positive (+) terminal of the dead battery.

- Step 2: Connect Positive (Red) Cable to Working Battery: - Action: Connect the opposite end of the red jumper wire to the positive (+) terminal of the functioning battery.

- Step 3: Connect Negative (Black) Cable to Working Battery: - Action: Connect one end of the black jumper wire to the negative (-) terminal of the functional battery.

- Step 4: Connect Negative (Black) Cable to Ground: - Action: Connect the opposite end of the black jumper wire to an unpainted metal surface on the dead automobile. This acts as the ground.

5. Start the Working Vehicle:

- Action: Start the operating car and let it run for a few minutes. This helps charge the dead battery.

6. Attempt to Start the Dead Vehicle:

- Action: Try to start the dead car. If it doesn't start, check the cable connections and verify they are secure.

7. Disconnect Cables:

- Step 1: Disconnect Black Cable from Ground: - Action: Remove the black jumper wire from the ground of the dead car.

- Step 2: Disconnect Black Cable from Working Battery: - Action: Remove the black jumper wire from the negative (-) connector of the functioning battery.

- Step 3: Disconnect Red Cable from Working Battery: - Action: Remove the red jumper connection from the positive (+) terminal of the functioning battery.

- Step 4: Disconnect Red Cable from Dead Battery: - Action: Remove the red jumper cord from the positive (+) connector of the previously dead battery.

8. Drive the Restarted Vehicle:

- Action: Drive the restarted car for at least 15–20 minutes to enable the alternator to replenish the battery.

Tips:

- Use Good-Quality Jumper Cables: - Invest in durable, well-insulated jumper cables for improved conductivity and safety.

- Follow Vehicle Manuals: - Refer to both vehicle manuals for particular jump-starting directions and precautions.

- Avoid Touching Cable Clamps Together: - Ensure the cable clamps don't contact each other during the jump-start operation to avoid sparking.

- Consider a Portable Jump Starter: - Portable jump starters are practical alternatives to utilizing another vehicle.

Chapter 9: Maintenance Schedules

Creating a personalized maintenance schedule

Maintaining a specific maintenance program for your car is vital for its lifetime and maximum performance. Here's a step-by-step tutorial to help you design a personalized maintenance plan:

1. Refer to the Owner's Manual:

- Action: Review your vehicle's owner's manual. It gives complete information on the manufacturer's suggested maintenance intervals and services.

2. Understand Your Driving Conditions:

- Action: Consider your regular driving circumstances. For example, numerous short journeys or stop-and-go traffic may necessitate more regular maintenance.

3. Identify Critical Maintenance Items:

- Common Maintenance Items:
  - Oil changes
  - Brake inspections - Tire rotations
  - Fluid checks and replacements
  - Air filter replacements

- Action: Identify key maintenance activities depending on your vehicle's requirements.

4. Create a Maintenance Calendar:

- Action: Use a calendar, whether digital or paper, to plan regular maintenance work. Note the suggested intervals for each service.

5. Set Mileage-Based Intervals:

- Action: Determine mileage-based intervals for services like as oil changes, tire rotations, and brake checks. Factor in the manufacturer's guidelines and your driving circumstances.

6. Consider Time-Based Intervals:

- Action: Some maintenance activities, such specific fluid changes or inspections, may be time-based rather than mileage-based. Note these periods in your schedule.

7. Prioritize Seasonal Maintenance:

- Action: Consider seasonal issues, such as winterizing your car or planning for summer road excursions. Schedule required maintenance work accordingly.

8. Include Inspection Points:

- Action: Incorporate frequent inspections into your routine. Check things like belts, hoses, and brake pads for indications of wear.

9. Budget for Major Services:

- Action: Plan for important services that occur at longer intervals, such as timing belt replacements. Budget for these costs in advance.

10. Set Reminders:

- Action: Use reminders on your phone or calendar app to tell you when it's time for scheduled maintenance.

11. Adapt Based on Vehicle Age:

- Action: As your vehicle ages, adapt your maintenance routine to meet any wear-and-tear concerns. Older automobiles may require more regular maintenance.

12. Document Services:

- Action: Keep a record of all accomplished maintenance activities, including the date, distance, and specifics of the service. This helps trace your vehicle's history.

13. Be Flexible:

- Action: Be flexible to altering your itinerary depending on unanticipated challenges or changes in driving habits.

14. Consult with Professionals:

- Action: If unclear about the appropriate maintenance routine, check with a skilled technician for specialized guidance.

15. Stay Informed:

- Action: Keep up-to-date with recalls or service bulletins relating to your car. This information may alter your maintenance schedule

Tracking maintenance and repairs

Keeping a careful record of your vehicle's maintenance and repairs is vital for its long-term health and your safety. Here's a strategy to help you successfully monitor maintenance and repairs:

1. Create a Maintenance Log:

- Action: Set up a separate maintenance log. This might be an actual notepad, a spreadsheet, or a dedicated software for auto maintenance.

2. Include Essential Information:

- Key Details: - Date of service - Mileage at the time of service - Type of service or repair performed - Parts replaced or repaired - Name of the service provider or mechanic - Cost of the service

- Action: Ensure your record captures all critical information for each maintenance or repair entry.

3. Regular Maintenance Entries:

- Common Maintenance Items:
    - Oil changes - Tire rotations
    - Fluid checks and replacements
    - Air filter changes
    - Brake inspections

- Action: Log each incident of regular maintenance, noting the date, miles, and specifics of the service.

4. Record Repairs:

- Action: Document all repairs, whether little or big. Include details regarding the issue, components changed, and the cost.

5. Attach Invoices and Receipts:

- Action: Keep a file or folder for invoices and receipts related to maintenance and repairs. Attach copies to the log entries.

6. Track Warranty Information:

- Action: Note whether any repairs are covered under warranties. Include data such as warranty durations and coverage.

7. Use Digital Tools:

- Action: Consider adopting digital tools like maintenance apps or software that can automate tracking and send reminders.

8. Set Reminders:

- Action: Schedule reminders for impending maintenance chores based on your log data.

9. Include DIY Projects:

- Action: If you undertake any do-it-yourself maintenance or repairs, record these actions with the same attention to detail.

10. Regularly Review Entries:

- Action: Periodically analyze your maintenance record to detect trends, possible problems, or places where expenses may be reduced.

11. Consult for Future Decisions:

- Action: Use your maintenance journal while making choices about selling, exchanging, or acquiring a car. A well-documented service history boosts your vehicle's worth.

12. Digital Vehicle History Services:

- Action: Explore digital car history services that may provide extra capabilities for monitoring maintenance and repairs.

13. Back Up Your Data:

- Action: Regularly back up your digital maintenance record, particularly if utilizing an app or program. Protecting your data guarantees you don't lose vital information.

14. Review for Recalls:

- Action: Use your maintenance record to check for any recalls or service advisories linked to your car.

15. Be Thorough with Descriptions:

- Action: When inputting information, be comprehensive with your descriptions. The more information you offer, the better for future reference.

Chapter 10: When to Call a Professional

Recognizing when a repair is beyond DIY

While doing do-it-yourself (DIY) automobile repairs may be satisfying, there are times when professional experience is required. Here are crucial clues to help you realize when a repair is beyond DIY:

1. Complexity of the Task:

- Indicator: If a repair includes sophisticated components, specialized equipment, or demands significant technical expertise, it's certainly beyond the reach of a DIY effort.

2. Safety Risks:

- Indicator: Repairs requiring safety-critical components like as brakes, airbags, or steering mechanisms should be done by specialists to guarantee optimal performance and safety.

3. Lack of Specialized Equipment:

- Indicator: If the repair involves specialist tools or equipment that you don't have access to, it's preferable to seek expert aid.

4. Electrical or Computer Issues:

- Indicator: Problems connected to the vehicle's electrical or computer systems generally necessitate specialized diagnostic instruments and experience, making them tough for DIYers.

5. Risk of Further Damage:

- Indicator: If you're unclear about the repair procedure and there's a great chance of inflicting more harm, it's best to see a specialist.

6. Fluid Leaks:

- Indicator: Repairs involving fluid leaks, particularly those linked to the gearbox, coolant, or braking system, may have catastrophic repercussions if not treated appropriately.

7. Warranty Considerations:

- Indicator: Performing some repairs on a new car could invalidate the warranty. Consult the car manufacturer or dealer to understand warranty consequences.

8. Time and Expertise:

- Indicator: If a repair involves substantial time and experience, particularly for individuals with minimal automotive understanding, it could be more effective to delegate the work to specialists.

9. Structural or Frame Damage:

- Indicator: Repairs involving structural or frame damage frequently need specific equipment and skills to maintain the vehicle's integrity.

10. Hybrid or Electric Vehicles:

- Indicator: Repairs on hybrid or electric cars entail high-voltage systems that necessitate specialist expertise and safety procedures, making them unsuitable for DIYers.

11. Air Conditioning System Repairs:

- Indicator: A/C system repairs sometimes include handling refrigerants, which needs certification and specific equipment owing to environmental requirements.

12. Excessive Noise or Vibration:

- Indicator: If there's a quick commencement of strange sounds or vibrations, it may signal a complicated condition that needs professional diagnosis and experience.

13. Lack of Confidence:

- Indicator: If you are doubtful, lack confidence, or are uncomfortable with the repair procedure, it's a good indicator to visit a professional technician.

14. Diagnostic Challenges:

- Indicator: If the issue is tough to identify, specialists with sophisticated diagnostic equipment and expertise may spot issues more precisely.

15. Fluid Changes in Modern Transmissions:

- Indicator: Many current automobiles have sealed gearboxes, and doing fluid changes without the correct tools might lead to issues.

Finding a reputable mechanic

Choosing a reliable technician is vital for the well-being of your car and your peace of mind.

Here's a guide to help you select a trustworthy and competent mechanic:

1. Ask for Recommendations:

- Action: Seek recommendations from friends, family, colleagues, or local community forums. Personal experiences are key markers of a mechanic's trustworthiness.

2. Check Online Reviews:

- Action: Look for internet reviews on platforms like Google, Yelp, or specialist review sites. Pay attention to both good and negative remarks to acquire a balanced view.

3. Verify Certifications:

- Action: Ensure that the technician is qualified by recognized organizations such as the National

Institute for Automotive Service Excellence (ASE). Certifications reflect a dedication to professionalism and competence.

4. Check Better Business Bureau (BBB) Ratings:

- Action: Visit the Better Business Bureau website to verify the mechanic's rating and any complaints made against them. A high rating suggests a dedication to client satisfaction.

5. Visit the Shop:

- Action: Take a visit to the mechanic's shop. A clean and well-organized building frequently displays professionalism and attention to detail.

6. Ask About Experience:

- Action: Inquire about the mechanic's expertise, particularly with your car make and model.

Experienced mechanics are more likely to identify and solve concerns properly.

7. Request Estimates:

- Action: Obtain written quotes for services or repairs. Compare these estimates with others in the region to guarantee fair pricing.

8. Understand Warranty Policies:

- Action: Clarify the warranty terms for both components and labor. Reputable technicians typically issue guarantees for their work.

9. Ask About Specializations:

- Action: whether your car needs specialist knowledge (e.g., hybrid systems, German imports), inquire whether the technician has competence in that area.

10. Inquire About Parts Quality:

- Action: Ask if the technician employs original equipment manufacturer (OEM) components or high-quality aftermarket parts. The quality of components may effect the lifespan of repairs.

11. Check for Professionalism:

- Action: Assess the professionalism of the personnel, including how they communicate and if they are eager to answer your queries.

12. Get a Second Opinion:

- Action: If a technician proposes significant repairs, consider seeking a second opinion to check the diagnosis and offered remedies.

13. Look for Transparency:

- Action: A good technician should be upfront about the diagnostic procedure, give clear explanations, and keep you informed throughout the repair.

14. Ask About Turnaround Time:

- Action: Inquire about the estimated turnaround time for repairs. A reputable technician sets realistic deadlines and discloses any delays quickly.

15. Trust Your Instincts:

- Action: Trust your intuition. If anything seems odd or if you're uncomfortable with the mechanic, try exploring for alternatives.

www.ingramcontent.com/pod-product-compliance
Lightning Source LLC
Chambersburg PA
CBHW052207220526
45471CB00004B/1847